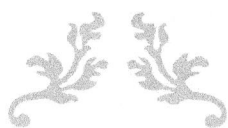

MEMORY TRAINING PRO

How to Use Advanced Learning Strategies to Learn Faster, Remember More and be More Productive

GAT
by Karen Chapman

Table of contents

INTRODUCTION   3

CHAPTER ONE: FOCUSING AND CREATING MEANING   4

CHAPTER TWO: "LINK METHOD"   5

CHAPTER THREE: „PEG METHOD"   7

CHAPTER FOUR: "LOCI SYSTEM"   8

CHAPTER FIVE: "JOURNEY SYSTEM"   10

CHAPTER SIX: "NUMBER SYSTEMS"   11

CHAPTER SEVEN: "ALPHABET TECHNIQUE"   13

CHAPTER EIGHT: "MAJOR SYSTEM"   14

CHAPTER NINE: "LANGUAGE. MAGNETIC MEMORY METHOD"   15

CHAPTER TEN: FORMING A STORY   16

CHAPTER ELEVEN: „CONNECTION TECHNIQUE"   17

CHAPTER TWELVE: "SNAPSHOT TECHNIQUE"   18

CHAPTER THIRTEEN: VISUALIZATION   19

CHAPTER FOURTEEN: „MNEMONIC MEMORY GAME"   20

   Blind' Jigsaw Puzzles   20

   Trivia Quizzes   20

   Pexeso: Matching Pairs   20

   Seize the Keywords   20

CHAPTER FIFTEEN: "CHUNKING METHOD"   21

CHAPTER SIXTEEN: "FIRST LETTER ASSOCIATION" TECHNIQUE   22

CONCLUSION 23

AUTHOR'S AFTERTHOUGHTS 24

COPYRIGHT 2017 BY KAREN CHAPMAN - ALL RIGHTS RESERVED. 25

Introduction

> We remember what we understand; we understand only what we pay attention to; we pay attention to what we want.
>
> Edward Bolles

How often do you forget people's names and their addresses? Are you often unable to recall an important fact or a phone number? These are common examples where a good memory is an important thing.

Memory is a human's ability to store, encode, keep in mind and further recall the information and experiences in the brain. It is the total amount of what we remember, the amazing ability that gives us a chance to adapt the learned facts, skills, habits, impressions. It is the so called „reservoir" of the things learned before or activity to keep in mind the necessary information; it can be understood also as the use of experience to influence or affect the current behavior. Memory as an important work skill can be improved through a variety of the different techniques. Whether it's keeping in mind the key statistics during meaningful talks, or impressing clients with your comprehensive knowledge of their product lines – your ability to remember is a major advantage and the memory techniques you chose to play the key role here.

The memory tool or stated differently „mnemonic" is a technique for remembering the information that is difficult to recall. The memory techniques are used to encode the information that isn't kept in our mind using the special methods. Here you will find the most common and easiest memory techniques that need only five-ten minutes a day to practice and you'll benefit from them. The more popular the "Loci" and "Link" methods, "Peg" and "Journey" systems, "Memory Palace technique", Visualization, Focusing, Snapshot techniques and also a wide variety of mnemonic memory games evolve, improve and interpret an information storage. Reviewing the following information according to the memory systems allows us to answer such questions as: How to make a right choice between rendered memory systems? How long does it usually take to apply the selected technique? What additional tools do we need? Which system is suitable for me? And dozens of other questions that may arise during further dipping into the freshest information. The more time you spend practicing the memory techniques, the quicker you'll see the results and more effective they will be. Spend ten to fifteen minutes a day looking through some memory technique while doing your daily routines. Remember that people with good memory are often seen as smart, knowledgeable, reliable and competent.

Chapter One: Focusing and Creating Meaning

It is common knowledge that the greatest power of the human's mind is an ability to focus on the things for a long run. Another matter, when you can't concentrate, everything seems more complicated and is time-consuming that you haven't expected. Maybe, you have looked to mend the concentration to organize a better work, to pass the exams or just to make the daily life easier. If you can't focus on something, think that it is just the way your brain works and that there's not so much you can do about it. It is clear, everyone can develop and improve the ability to concentrate, focus despite distractions nourishing your brain and using the simplest concentration techniques.

It's needless to say that your concentration depends on your memory, once you improve your concentration you improve your memory. To concentrate easily means to focus tensely.

If you don't concentrate, you won't be able to keep in mind the facts or bring back to memory the information you have memorized beforehand. Concentration is a mental skill that could be exercised and developed. There are some ways to enhance your concentration: First, raise your brain's natural ability to concentrate, i.e. focus on the most important things. In other words, increase your attention. Second, arrange the surrounding environment to make your concentration easier. To improve your concentration takes a little time and effort, but that's worth it.

Chapter Two: "Link Method"

The „Link Method" dominates among the advanced memory tools that are available once a person understands the concept of both substitute words and visual association. The method is called the "Link" because of the using it for information connection in a person's mind with the links being turned into a chain. Put the matter another way, it is hooking the information together one is trying to learn. It seems like the series of dominoes, so one can remember items that may not be related by linking them one to another. The most-effective substantial memory training methods that involve the "Link Method" also, are applied for the visualization and association technique. This method is a valid mean to change the abstract facts into the mental pictures making them easy to remember. The brains of the human beings can remember the visual images (pictures) much easier that any abstract information, for the very reason a person can usually visualize in his/her mind a certain place he/she may have visited even if it is impossible to remember the name of this certain place.

The steps for applying the "Link method" are the following:

1. The first thing - one should imagine some simple, memorable image that can represent the kind of list a person want to remember. This image should include the first item of the list that will be such a header for the rest list;
2. One should think of another simple, memorable image that joins, in other words links the first already existed item to the second one;
3. One thinks of another new image that joins the second item to the third one;
4. One should think of an image that may link the third item to the fourth one and so on, creating the mental images one after another for the residual items in the list.

This method of memorizing a list may seem strange if one has never tried this technique earlier. The next example can explain how to use the „Link method": we have a list of following spontaneous picked words: table-pen-locket-umbrella-salt. Following the steps given above, we should firstly link "table" to "pen". It is possible that a person thinks about a table that may have the shape of a pan or a person can see a pen on the table. Then, according to the provided pattern, a person has to connect word "pen" with "locket". Imagine that a person visualizes a pen with a locket or a pen that may hang from somebody's neck like a locket. Continuing the prescribed rule, a person links word "locket" to the word "umbrella", where another locket may be shaped like an umbrella. The other meaning may be an outdoor market selling knickknacks where the word "lockets" may be written in large letters on the umbrella attracting somebody's attention. Finally, one should link the word "umbrella" to "salt," as an example, picturing a famous Morton® Salt container, where a girl protects herself from the strong rain of salt using the umbrella. The linked list looks like this: thoughts of a „table" remind about a „pen", „pen" set up a mechanism of remembering of "locket" that will lead to the word "umbrella" and finally the "umbrella" helps to recall "salt". The "Link Method" of the memory training is also used to memorize a shopping list. Let us suppose, someone has to buy food at the food market, but unfortunately doesn't have any piece of paper to write down the whole list that contains: six red apples, loaf of bread, a carton of juice, bar of foamy soap, one pair of blue socks, a packet of chocolate biscuits. Beginning with the creation of a mental image of the grocery list, one may imagine a usual shopping cart that could be found at any food market. Basing on that fact that people's brain remember the unusual, not the ordinary, one may remember six red apples been smashed onto the shopping cart. The vivid, racy pictures are better for memorizing than the usual dull scenes, that's why one should „picture" as crazy as only possible. Just imagine the apple juice running down the shopping cart so that one can see the bright red color of six apples. So, the first link in a chain will be six red apples smashed onto a shopping cart. Continuing to make a memory link it is worth noting, one has created a simple mental image joining the concept of the „grocery shopping list" with "six red apples". The next position in the shopping list takes the second item – a large loaf of bread. The manner of memorizing the items continues through the rest of the shopping list. Thinking about some memorable image that may connect the first item – six red apples, the mental image may include an action, that will be more memorable for a brain. Let's imagine a strong rain of red apples that smash down on the large loaf of bread. In such a way, trying to make the objects that have to be remembered, rather large, makes them memorable. For example, the loaf of bread can be imagined as gigantic, like a big house. Following the grocery shopping list the next, third item that should be remembered is a carton of juice. It may be a giant carton of juice kicking a large loaf of bread like a football. The most important thing here: the more funny, strange is the image, the easier it will be to memorize. It could be even a team of juice cartons kicking a large loaf of bread running down the field. The next item in a shopping list is a bar of foamy soap that one should link to the previous item of a carton of juice. The link that could be thought of, is a juice pouring from the carton, but instead of it, there is a foamy soap! Would you like to drink it? No. Remember, the item from the shopping list are common, but the combinations are definitely not. This is the link image that the brain can't forget. Each of the images must connect, in other words, link to the next image. The fifth item on the shopping list is a pair of blue socks, that demands an image connecting both items of the blue socks and a bar of foamy soap for creation of the next link. An image of washing blue socks in a foamy soap wouldn't work because this procedure is not unusual. One needs a bizarre, strange, funny image. Let's imagine a „soap man" putting on a pair of blue socks. And the final item in the shopping list will be a packet of chocolate biscuits, that should be linked to an image of a pair of blue socks. Why not put the chocolate biscuits into the blue socks? But it is impossible to go in such a way, the broken biscuits hurt the feet! And now, when somebody is already at a grocery store, it is easy to remember the thing one need to buy: thinking of a shopping cart reminds you about first image – six red apples; the apples raining down from the sky on a large loaf ob bread; the giant carton of juice is kicking a large loaf of bread like a football; the pouring juice from the carton on a bar of foamy soap reminds about the next item of a foamy soap; suddenly appears the next image of a soap man putting on a pair of blue socks, and finally – the chocolate biscuits in the blue socks that make the walking very uncomfortable! The „Link method" of memory training allows remembering items for some days, even if it isn't used. It can be also used an infinite number of times. The retention of the list will be more effective in a comparison with the use of visualization technique.

Chapter three: „Peg Method"

The "Peg System" is one of the most useful techniques for memory training. The "Peg Method" uses visual imagery to provide a "peg" or a "hook" to hang (or rather to associate) somebody's memories. The "Link Method" is excellent for remembering lists in sequence, but it doesn't allow the easiest way to recall, for example, the 8th or the 9th item in the list. One should have to start from the very beginning of the link and count mentally through the imaginations and associations until the 8th item will be reached. The "Peg Method" provides an ability to remember the numerical position of an item, namely in a list in a sequence. A "peg" is a so called „mental hook" on which one should "hang" the necessary information. The hook is a reminder for retrieving the information mentally. The "Peg Method" provides also a system – *the "Rhyming Peg"* where one should use a sound-a-like word for construction of the number of the pegs. It's not necessary to use rhyming words to form the "pegs", but the numerous studies have shown that the recall was substantially improved by doing this way. The "Rhyming Peg" method is appropriate for the short lists and available for the lists containing more than 10 items. Beginning with rhyming or associating a sound-a-like word with the number from 1 to 10. So, one may use the own words (associations): 1 – bun; 2 - shoe; 3 - tree; 4 – door; 5 – hive; 6 – sticks; 7 – heaven; 8 – gate; 9 – vine; 10 – pen. Continue closing the eyes, saying the number and trying to make a blazing, vivid mental image of the object you rhyme. Doing this several times and reviewing the images over the next days, the images will be fixed into the long-term memory of a human being. But what if somebody forgets a "peg" word? The other easy way to join the "peg" words to the memory and to make the way of their recalling easy, is to create a short story using the "Link method" that we are already familiar with. As soon as the "peg" words are encoded with the corresponding numerical counterpart, it will be easier to make an association. The items that have to be remembered are associated with the "peg" word image by means of elaboration skills and the encoding. This time it is possible both to recall the list and to know the exact position (number) of each item in the given list. Just to check the efficiency of the "Peg method" one may ask someone else for help providing him/her with a list of items with special numbers. Ask them to quiz you to a numerical position of the items and vice versa. Practicing the "Rhyming Peg" technique makes clear - it is one of the efficient and simplest ways to memorize the whole list of the items and their special position in a list.

Chapter Four: "Loci System"

One of the oldest memory training methods is called „Locy Method" („loci" is a plural form of the word „locus", it means „place" or "location"). The main idea of this method is a hypothesis that a person can better remember those places that are already known. This technique was introduced more than 2000 years ago, and was widely used by the Romans and Greeks to memorize and to launch into a harangue that could last for hours. Nowadays, most part of the audience could also be impressed by a speaker who is talking without referring to some notes, who is giving an exciting speech from memory conveying authority and competence. Could you imagine yourself giving a half an hour speech to the point without using any notes? The "Loci Method" is appreciably a visual filling technique that allows us to recall and memorize an imagined unlimited number of the items in a special fixed order. Each of the locations is applied like a „hook" to which one should connect an item or something else that must be remembered. This is performed by creating a scene or an image in mind, in which the to-be remembered item or location interact. The sequence is provided by determining a specific, precise journey with the distinct locations along all the route a person is familiar with. Like any journey, when a person needs to arrange a fixed point to start the route. The method of "Loci" works well if a person is good at visualizing of things. The working plan of the method is following: firstly, we think of a place we know well, as for example our own house. Then, we try to visualize some series of the locations in the place in a logical order. For example, you describe the way you usually take in your own house to come from the front door to the back door. Beginning at the front door, you go through the large hall, then turn into the living room, proceed into the kitchen, and so on basing on the number of the rooms in your house/flat. As you enter each of the rooms, you move consistently and logically in a same direction, from one side of the room to another. Each room could serve you as a location where you may place the item you wish to remember (one item at one of the locations). Here is a simple shopping list you need to remember: body wash, apples, buns for hot dogs, ketchup. As you visualize your own house, try to imagine you water the front door with the body wash. But don't imagine the word „body wash", just really see as you press the tube all over the front door. You may try to imagine the smell of the body wash. Do you like lavender? And now, you open the door, enter your large hall and try to imagine a big red apple rolling down the steps in the front hall. You turn further into the living room and visualize a big hot dog in a tasty bun lounging by the fireplace in a cowboy hat. Now you enter the kitchen where a bottle of a bitter ketchup dressed in an old-fashioned uniform is serving up. After all the items on the list have been placed in the house, try to remember the shopping list and goods you need to buy; all you need to do is just to visualize the front door of your own house.

This method can be used to remember the important points in a speech, names of meetings, events or people, everything you wish to keep in your mind. This technique works rather well because it allows changing the manner you remember. The familiar locations of your own flat/house, which are organized in a familiar order, or other places prompt a person about things. The „Loci Method" of memory training can be applied by adding any buildings you know very well: a mall, your office building, your garden or a friend's house -- any place you know well. The one important thing in this technique – is how different are places one from another. It's of great importance to have an association between a location and an item, the so called „interaction" just to make each of the locations, memorable and different

The both methods – the "Link method" and the "Loci method" allow us to remember the special items on a list, but neither let us appoint just a single particular item. This time, if you have to find, for example, the tenth item applying the "Link method", you should process all the first nine items to get to the tenth one.

Chapter Five: "Journey System"

The "Journey Technique" is an excellent way of a simple and successful memory training. Thanks to its simplicity every child could be taught the exercising of this memory technique. Instead of difficulties in remembering dates, names, famous people and all the rest information, memorizing could be carried out like a game. This method would appoint both kids and adults for a lifetime of exciting memory work that is more enjoyable and not boring.

The "Journey Method" of memory training works in a simple way: if a person can find a way around a place in the imagination, he/she can leave things along the way that means they'll be there next time when a person takes this journey again. This method seems to be the easiest one, doesn't it? What does a memory journey consist of? It could be something like strolling through the own house that you know well already. Let's imagine strolling in the neighborhood, you know the shops, streets, buildings, parks, statutes and other kinds of things that you have already visited. Each of them can serve like a „hook" (do you remember the "Peg Method"?) on which one can "hang" something you need to remember. Is it a big railway station near you? Think of all the „stages" ("Loci Method") that could be recited. Remember all the locations that could be found there. The following guidelines should be used in the "Journey method":

1. You should choose a memory journey along the route you know already. In case, if the route isn't familiar enough, examine it until you become familiar with it;

2. You should determine the most interesting and distinct locations that will be used further as files for keeping the necessary information;

3. You should always look at a location from the same perspective;

4. You should choose a specific characteristic at each station as a "peg". It allows for a more sustainable link between the information and the location to be memorized. Based on what a person wants to memorize, he/she recedes from this guideline.

If a person needs to memorize things constantly, go off on a long-term journey, and make it more complying to the subject matter. Let's imagine you're working in a museum. You could use the museum for your memory journey, and conceptually place selected items at the key places to give an idea, a cue to your memory towards the special things you need to memorize. Examine the place carefully and then get used to mentally placing things at the front door, the halls, the galleries, in the wings, at the box office and so on. Do you have any art gallery or a museum in the area you live in? The buildings like these would be ideal locations for your memory journeys. If somebody wants to change this into long-term memories, one should revisit these locations – this is a great method of improving your memory. It will take a minute or two to stroll along the house or other place you have already chosen, viewing all the things in their respective places, and using your perception to really experience them. But don't gallop through your journey, speed is not the most important thing here. Take a good look at things, touch them and walk around them. Is it something that can be really smelled? Imagine the smell, breathe it deeply and enjoy. Touch the items, when it is possible to move them slightly – do it! Feel its consistence and weight. This will all be essential in your real memory. Once the journey is already arranged, you can revisit it for some times once or twice the same day or the next day. Keep in mind to get back later, a week or a month later, after this time you will have reinforced this memory. Memories created in such a way, can become constant without a lot of effort.

Chapter six: "Number Systems"

Have you ever tried to remember numbers? It is rather hard, isn't it? If you have a wish, to form a great memory, you have to learn a way to memorize the numbers. The "Phonetic number system" was developed over 250 years ago and keeps one of the leading places among the memory techniques using associations and imaginations. How do you think, why it is important to remember numbers? Try to imagine, your chef assigns a task for you to give a sales presentation of the court of directors of the company. The presentation should include sales statistics and figures, so, you have to meddle with papers or spit out all the numbers from your memory. And what about if you need to remember the street addresses, phone extensions, passwords, product codes, mathematical constants or something like this. Would you like to try the "Phonetic number system"? This system is just an alternative to the alphabet that changes numbers to letters. Just an example: if you have to memorize the simple phrase, "Five score and seven years ago", I think, you could easily do it. You may repeat this phrase every time during a day, it is not so difficult for the memory - even though that phrase consists of more than twenty letters and spaces in a specific order. But what about to memorize such a number, like "7658 34531 887 54365 23678 963". It is pretty impossible, right? But both the number and the phrase contain 25 characters and what's the difference? The difference between the first phrase that is simple to memorize and the second is caused by the case from the chunking effect of the alphabet when the letters can be used to form the words. The letters are arranged into words, and words are arranged into expressive phrases. A shocking, powerful trick for memorizing the numbers is found in a way to transform the numbers to letters. After that you could take any 25-30-digit number and try to turn this into a series of phrases or words. The "Phonetic number system" works in following order:

An example of the conversion table. If you want to use the "Phonetic number system" for memorizing numbers, you have to remember this table firstly. By the way, it's not so hard as it looks. Let's consider an example: 1 = is the "d" or "t" sound; 2 = the "n" sound; 3 = "m"; 4 = "r"; 5 = "l"; 6 = is the "j" sound/ "ch", or "sh" sound; 7 = is the "k"/"g" sound; 8 = is the "f'/"v" sound; 9 = is the "p"/"b" sound; 0 = is the "z"/"s" sound. The following mnemonic recommendations make it easier to memorize the table. Be attentive, when more than one letter forms an association with a number, the letters can sound similar (for example, "b" and "p" for "9"). 1 - a typewritten "d" or "t" has only 1 down stroke; 2 - a typewritten "n" has 2 down strokes; 3 - a typewritten "m" has 3 down strokes; 4 - the number ends with the "r"; 5 - hold out your left hand, palms outward, thumb out at a ninety degree angle - the five fingers form so called "L" shape; 6 - a "J" looks like a backward of 6; 7 - a "K" is made from two back-to-back 7's; 8 - a lowercase, written „f" looks like an 8; 9 – an upside-down „b" or backward "p"; 0 - the word "zero" begins with "z". Try to memorize this list during 10 minutes of practice. Be attentive, all the rendered letter equivalents are consonants. The vowels are not appointed to a number in the method of the "Phonetic number system". The consonants "w", "h", or "y" (w-h-y) are also absent. And now, using the above mentioned technique, try to explore an example of memorizing a 6-Digit passcode. Imagine, you are working at a company where each of the employee has a personal, unique passcode that must be punched on a special keypad entering the company's main office. The security service changes the passcode within a given period and doesn't prefer the employees commit it to paper. You have been given a new code recently: 954392. Memorizing these numbers is rather difficult, but we are already familiar with the "Phonetic Number memory system", this number could be transformed into one or a couple of words: 9 = b, p / 5 = l / 4 = means „r" / 3 = m / 9 = once again = b, p / 2 = n. If you remember an equivalent table above, an easy-to-remember little phrase can come to your mind: "BoiL RuM PaN" or maybe "PooL RooM BuN" or any other phrase is possible. Take this simple phrase and think quickly of some visual association to memorize it. For example, a big hot dog in a tasty bun playing tennis in a pool hall. Practice, each time when the password is required and you should press the buttons on the keypad to enter the main office, think of a „hot dog bun playing tennis in a pool hall". It is clear that the "Number system technique" really works, it is not only a kind of fun. Put the so called „letter transformation" into a practice with the long numbers you need to memorize!

Chapter seven: "Alphabet Technique"

The "Alphabet technique" is similar to the "Peg method" (see chapter three) and it is a little more advanced than the "Number system" (see the previous chapter six). This system is rather good for remembering the long lists of the items you need in a special order when the missing items on the list can be easily disclosed. A work-manual how to use this technique: this method works by coupling together, i.e. associating the images cued and represented by the letters of the alphabet together with the images that perform the items to be kept in mind. The choice of the images that represent the letters isn't rested upon the primary character of the letter name. The images are picked out phonetically - i.e. the name of the letter corresponds to the sound of the first syllable of the image word. For example: the letter „c" and the word „sea"; the letter "g" and the word "jeans". You should create an „image scheme", for example: „b" = bee; „I" = eye; „k" = „cake" and so on. Ideally, it should be the first image which springs to your mind. Once person distinctly visualized the images and have joined them to their basic letters, it is possible to associate them with the information to be kept in mind. Try to imagine, you have to remember a special list of characters from the TV-series YuGiOh: A (Ace) – Kaaba – an Ace that is printed on a Kite being flown outside a Bar; B (Bee) – Joey – a Bee stings a baby kangaroo; C (See- saw) – Mai – a giant Mite poised on the seat of a seesaw; D (Demon) – YuGi – a demon that eats YoGhurt. Finally, the Alphabet method links the items to be kept in mind with the images of the letters from A to Z. This technique allows you to stick in memory a medium-sized list in a certain order. By fixing the items to be retained in the memory to the letters of the alphabet, you always know, if you have forgotten the items, and know the cues to use to launch their recall.

The „Alphabet Technique" is the most complicated system of the memory techniques which requires the longer arrangements and is more difficult to encode than the "Number" or other systems, we have already spoken about. This way of the storing the information in your brain allows to remember and code a list consisting of more than 20 items. Probably, you should use more effective, simple "Peg system" in connection with other techniques that the "Alphabet system". It is your personal choice!

Chapter Eight: "Major System"

After learning the basic "Loci system" and other simple techniques, exercising a perfect and uncommon association, it's time for the benefit performance of the "Major System". The majority of the mental athletes uses this powerful tool with a great pleasure. This system is based on the phonetic technique that allows you to keep in mind all the essential images without effort. The numbers are really hard to remember, nothing seems to make them memorable. But there is one tool that will blow your mind! The "Major system" is a kind of a phonetic technique used to transform the numbers into the words. The system begins its function with transferring digits into the consonant sounds, adding the vowels before, between and after those letters. The phonetic rules are easy to remember because they are mnemonic and this will help you to restore the images in your head even assuming that you haven't learned the whole set of 100 images yet. Practicing this method allows you to become strong associations beginning from 0 to 9. All you have to do is keeping in mind the number that is associated with the sounds which are easy to link together to make the words. You can also join the sounds together for making longer words, as long as you want – and this can make for an amazing degree of flexibility and variation. The key target of the "Major system" is to convert numbers to images and vice-versa. The 10-item mnemonic table shows how to transform the digits from 0 to 9 into the appropriate sounds which will be finally applied to make words. It takes 20 minutes usually to learn the mnemonics and to master them fully. We have already got acquainted with the numbers and correspondent letter in chapter six: "Number Method". Let's review and refresh them again! Here they are: 0 = „s, z, soft c" ("z" is the first letter of the word „zero"); 1 = d, t, th ("t" & "d" have one down stroke, sound similar); 2 = n (two down strokes); 3 = m (number „3" lying on its side, has three down strokes); 4 = r (the last letter of „four"); 5 = L (the Roman numeral for 50); 6 = j, sh, ch, dg, zh, soft g ("6" looks like a whistle); 7 = k, hard c, g, q, qu („K" contains two 7's back to back, on their sides); 8 = v, f (be attentive how your teeth touch lips for both letters; „v" as in a V8 motor); 9 = p, b (p – is a mirror image of 9; b – resembles with 9 rolled around). The vowel sounds as the sounds „w", „h", „y" can be used at any position without changing the word's number value. Let's consider number „42" as an example: following our mnemonic table, the single-digits in the number 42 are transformed to the corresponding letters "r" and "n". Now we should create any word that dawns on with the letters "r" and "n". It is necessary to fill the gaps between the letters by the means of the '"neutral" elements (Remember: The vowel sounds as the sounds „w", „h", „y" can be used at any position without changing the word's number value). The first word that is at the top of my head is „rain", so, the number 42 gets encoded as *rain*. The process of conversion seems to be a little burdensome and slow at the first, the practice turns this around! Just a couple more notes to keep in mind: Please, remember, the conversions are strictly phonetic, i.e. they are based on how the word sounds (not how the word is spelled); if a word has double letters, they have to be accounted for a one sound, you count only one sound (for example: the „r" sound in the word "lorry" counts as only one). While working with the words, you should choose those that are easy for you to visualize; concrete nouns, for instance, animals or objects, always work better than verbs, abstract nouns or adjectives.

Chapter nine: "Language. Magnetic Memory method"

Have you ever learned/ tried to learn foreign languages? Well, you should know that learning a new language is more difficult than improving the one you already know. Knowing more languages gives a chance to become a more valuable worker, to travel to different places of our world and communicate easily. Acquiring a quite new vocabulary that is different from the native language can overload the traditional memory and be time-consuming. The "Magnetic memory method" as one of the means how to teach training the brain by grasping the new words, makes learning of a new language easy and fun. The Magnetic memory technique teaches you how to use various elements of your daily routine and how can the world you live in be turned into a "Memory Palace". The "Magnetic Memory method" is used for systematically remembering anything you want! So, a so called "Memory Palace" is our mental construct applied to keep and to restore information to be remembered. A "Memory Palace" is much more powerful than any other single mnemonic by the reason of possible keeping and restoring of another memory strategies inside of "Memory Palace". This "Palace" is based on the real locations – your home, places where you have studied (schools, universities), offices where you have worked, places in the immediate neighborhood, etc. There are endless options, you can use literally any place with which you are familiar to keep the information in mind. Do you remember our „Journey system" (chapter five) when we have "hooked" the words from our shopping list of the locations in your house, office or somewhere else? We leave words and phrases we need to remember in the station or at the rooms of the "Memory Palace" and then we will pick them up when we will go through the rooms of our house/ flat, office building or something else. Once we have chosen the "Memory Palace", we draw a floor plan. Before we number our stations, we form a linear path through this floor plan. The technique of the "Memory Palace" works best when you don't cross your own path. Don't squeeze each possible situation into your first place, include the trite locations like a kitchen, bathroom, living room, bedroom and an entry point (It's better to draw the picture with locations). Don't forget to use only the building you are familiar with. Then make a top-down list of the stations in your palace in a linear order. And now let's place the words using three classic principles: 1. Paying attention to special phrases and words; 2. Encrypting the sounds and the sense of information with the use of action and imagery (each phrase or word becomes memorable); 3 Decoding actions and image for further moving words and phrases into the long-term memory. Remember: encoding your information needs creation of bright, colorful, large images as at all previous methods! Don't be a cracking bore!

Chapter ten: Forming a Story

The stories are probably as old as a language, and it seems we are deeply attached to them. The most part of our means of understanding the world are narratives of one or another form, whether conceptual metaphors, serious scientific stories or the kind of tales we use to acknowledge our choices in life.

The recent researches have asserted that our memory works much better by creating the vivid images and stories whenever we are learning. The psychologists strongly believe that getting students to form little stories would enhance them to remember more easily what they are studying. It mustn't be a large story, just two-three lines of necessary information, including the traditional beginning, the development of the action (middle) and the ending. A personalized story is still one of the best memory techniques! Do you remember the "Alphabet technique" and the "Number system" where you had to remember the encoding numbers and letters? You don't need this here at all! The effectiveness of this memory technique is that even abstract concepts can be kept in your mind. Of course, as a rule, the memory techniques work best if you can use easily visualized „things" unlike the abstract concepts. The last searches showed that generating a "nano-narrative" for even an abstract concept could enhance the student's recall of the material when tested some days later. When a person is including himself/herself as a character in the story, it makes the so-called „anchors", i.e. points in the memory which could identify the images, the words you need to remember. The story technique is based on the "Link method" (see Chapter two) that is the easiest to use and to understand. The "Story system" links the images together into a short story and keep events in a logical order. This allows you to improve the ability to remember essential information once you forget the sequence of images.

Chapter eleven: „Connection Technique"

"Connection technique" is a type of mnemonic where the information to be remembered is connected to something you already know. Most of the memory problems are due to our lack of attention, so the information never gets encoded (processed in a meaningful way) and never makes it to storage. Pay attention to what you know already, it will be your first step toward improving the memory skills. Then a simple technique of "Connection" helps you to handle and process the information so it stands a better chance of making it to storage for resumption. Have you ever found yourself in a room and couldn't remember why you were coming there? What do you usually do in such a situation? In most cases, you try to remember, i.e. „connect" things to images, data, actions that you are familiar with. „Moving" backwards you connect one thing or action to another to remember the goal of visiting the room. There is one more example here: How to remember the directions of the latitude and longitude? It couldn't be any easier. You know that the lines on a globe run North and South and are long, so, that complies with the LONGitude. Another Connection technique specifies that there is the letter „N" in LO<u>N</u>Gitude and an "N" in <u>N</u>orth. Latitude lines run east to west, then because there is no letter "n" in latitude.

Chapter twelve: "Snapshot Technique"

Never want to forget that beautiful sunset, gorgeous rainbow or other special events in your life? Using the brain and eyes like a camera, consciously make mental snapshots while it's happening. By involving a full concentration on the present moment this method provides the comprising of stronger, clearer memories of important scenes. If your lousy episodic memory is unreliable, take heart! Rely on "snapshot technique" if you want to increase your memory.

Selecting the best photos to the album, recording a video of an event or describing this event in a journal entry is the important way to store memories. The physical memory aid isn't geared to be carried around. A clear image in our memory, recalling in a moment, is obviously priceless. The "episodic" memory is the memory of events, a kind of verbal (declarative) memory, recognized by the cognitive neuroscience. The episodic memory includes memory of times, places, emotions, forming an association with the events that have taken place in your life. When you remember your wedding ceremony or what you had for dinner this afternoon, your episodic memory is at work. Some of us bring to mind the details of the events easily, whether it is a recent business meeting or the vacation of the previous year. The mental images stick in their brain strongly. The rest of people try to recall these memories without success. We often feel ourselves shyly admitting that we don't remember some significant events, that's why this clever memory technique may be for you! Taking mental snapshots doesn't involve the special tools: just your brain, eyes, an intention to concentrate, and a couple steps that imitate the use of a camera. The bottom line is to create a picturesque snapshot that is based on a clear episodic memory attention. One of the matters of poor and badly formed memories is an absence of the concentrating carefully at what's happening. Just imagine, you are attending the wedding of your friends and you want to remember the moment of oath in eternal love and fidelity to each other. You should concentrate keenly on a happy couple and try to make a mental photo. Realizing a situation you will connect deeply enough this special and important moment. So, to take a snapshot you should follow the same steps as if you were making a picture with a camera: your eyelids are like a camera's shutter. Follow these steps: Gaze at a scene with a great focus, attention and concentration. Realize how important is to remember this moment and the details. Chase all the trifles, colors, dresses, lighting in the room, smell, etc. Slowly blink your eyes and click your brain's camera shutter. Have you heard the „click" of your camera? You may review the scene in the mind's eye, just close the eyes for a moment and conjure up an imagine. The creation of a permanent and clear snapshot is based on an intense focus that requires a mental endeavor. Do not take dozens of snapshots at once, concentrate on a few special important moments that convey the meaning of the event. Creation of the mental snapshots is easier for persons who visualize well, from birth or who has developed the ability through practice.

Chapter thirteen: Visualization

The best memory technique used by the professional memory performers is the method of visualization. It is a fun memory trick that anyone can easily learn. Do you have problems remembering facts related to your job or school, work, maybe people's names or any other essential information? The solution is to get acquainted with the technique of visualization and a personal habit you use every day. The application of the visualization method of memorization is almost unlimited; it is used to help for recall foreign vocabulary, recall the definitions, lists, stories, poems, speeches and dialogues. The visualization method could be used together with the image-based memory techniques, to remember information well and easily. This method takes on board an amazing fact about human's memory: the images are remembered better than written or verbal information. For example, one can easily see imaginatively the homes that one has lived in during the life, or a school (including the interior rooms) which one has attended, while the phone numbers and addresses might be hard to remember. The visualization technique allows to convert the abstract information into the mental pictures that are easy to remember. These images are the so-called „mental hooks" enabling you to recover information from the long-term memory. The images are easier to memorize, creating the mental pictures is a powerful way to focus the mind. The visualization method takes a few minutes. Imagine, you have to remember the following list of items – banana, monkey, phone, apple. How would you do it? The first one is very easy – imagine banana and a monkey, who is eating banana. Keep in mind this image and really focus on it for a minute before doing the next one: „monkey – phone". To memorize this combination you should imagine a monkey and a phone or a monkey talking by phone about parenting. The more extraordinary the image - the better. Next, try to imagine a phone that looks like an apple, or a man stepping on and crushing it suddenly. Take a couple minutes doing this one, and repeat the previous ones before you move onto the next image. Then try to restore your list and you will find it comes amazingly easily.

Some researchers believe we never forget anything. In most cases, the reason a person can't remember is that the necessary information couldn't be found in our brains. We haven't become it a habit to create the mental "hooks" (the mental images) that we need to grasp and drag out the information. The practicing visualization memory technique gets a good chance to create the mental "hooks".

Chapter fourteen: „Mnemonic Memory Game"

Have you ever looked up your personal data and repeated it over and over and reviewed, until you dialled it correctly? Do the previous memory techniques seem to you a little bit boring? You can't commit the long number to your long-term memory? During the working process the memory is good at recalling of bits of the information and can hold a few pieces of information only for a short period of time. There are, of course, many memory techniques we have already had a look at. Let's have a bit of fun "working out" with memory games. There are many mnemonic memory games for keeping information in your brain, here some of them:

Blind' Jigsaw Puzzles

One of the funny ways to strengthen your memory is the good, old-fashioned "Jigsaw puzzle". Playing it „blind" means without spying and referring back to a picture on the box! First, look at the picture of the completed puzzle, take a few minutes to fix it in your memory. Second, mix up all the pieces of the "Jigsaw puzzle". Next, work to bring them down together without looking at the already completed picture of the "Jigsaw puzzle" (until you are done).

Trivia Quizzes

One of the ways to improve how well you bring back the information to memory is to play "Trivia quizzes". The word „trivia" can be about anything you wish – histories, movies or your specific business. Each person represents a list of the questions (with answers!), then a „quiz master" takes the questions from each person's list in turn. Playing with a new set of „trivia questions", you rely on a recalling of the prior knowledge and personal experience to find the proper answers. If you play with the same questions some days or even weeks later, you will also rest upon your memory of playing the game last time. Both the new and the repeated questions are good for building and strengthening your memory skills.

Pexeso: Matching Pairs

The game „Pexeso" involves the matching pairs of like cards from a large group, whilst one of each group is bosomed. Playing Pexeso provides use of a set of cards that includes pairs of the numbers or pictures. Start by spreading of 24 cards, making certain all the 24 cards consist of 12 matched pairs. Once face down, mix all the cards around so that you do not guess where any single card is hidden. Turn one card over at a time, take a look at the objects or numbers and turn it face down again. You can renew this process until you turn over the card that matches to that one you have turned over earlier. Now find the card's „yokemate" by recalling from earlier where it is located. Once you find a matched pair, retrieved them from the group.

Seize the Keywords

This game consists of reading a 10-line story (the story is chosen previously) and memorizing the order in which all the verbs (subjects, adverbs and so on) appear. Being attentive enough and understanding the story, makes it rather easy to keep in mind the logical order of the verbs (subjects, adverbs, etc.). Then, all the words must be listed at random and you will be asked to place them in the correct order, as found in the text.

Chapter fifteen: "Chunking method"

The „chunking" is a term addressing to the procedure of taking individual pieces of information (i.e. „chunks") and clustering them into larger entires. While grouping every piece into a larger whole, one could improve the amount of information you can remember. The most common example of chunking takes place with phone numbers. For example, a phone number sequence of 5-6-1-1-3-2-5 would be chunked into 561-1325. By separating incomparable individual elements into larger series, the information becomes easier to renew and retain. This is due mostly to how limited and narrow our short-term memory can be.

How to use the „chunking technique" to memorize things?

Once you try to remember the items from the list, start to form smaller groups. If it is a list of vocabulary words you might create small blocks of the words that are similar or related with one another. This shopping list might be broken down into smaller blocks related to whether the items on the list are fruits, vegetables, drinks or grains. The "chunking technique" can be used as an everyday memory booster. The effective "chunking technique" involves: looking for connections (look for means to join units to each other in a meaningful way; pay attention to what do the items share in general?); associate (connect the groups of the items to things from your memory, make them memorable!); incorporate other memory techniques (for example, once you have to remember the shopping list, create a key-word of the capital letters: bananas, eggs, nuts, tea = BENT. For this method see chapter sixteen).

Chapter sixteen: "First letter association" technique

The "First Letter Association" method is a kind of memory training techniques, where you should take the first letter of each word in a list of the words you wish to memorize and make a word or a couple of words associating to it. The most popular examples of this technique in our daily life are acronyms and abbreviations. But why the "First letter association" technique is so popular? There are some reasons: 1 – this technique always gives us keys = the first letter gives us a prompt performing, a big boost to our memory; 2 – this technique forces us to focus on attention and assigns a specific meaning in what we are learning at the moment; 3 – this method allows us to keep in mind more by memorizing less. While using this technique we abbreviate the material we have to remember. This method takes the first letter of each item or word on a list and forms an expressive word from those letters, that shortens the amount of the information you need to commit to memory in a substantial way. For example: How to remember the names of the five Great Lakes? They are Huron, Ontario, Michigan, Erie and Superior. Try to use the first letter association technique and you'll get HOMES. The same example is for the abbreviations you already know – USA = United States of America.

Conclusion

Our memory is something we deal with every day of our life, even when it seems like we are not currently using it. Memory allows us to keep in mind skills that we have already learned or receive information back that is remaining in the brain. Memory organizes information in a such way we can apply it in the proper context and use it in the present situation we are involved in. The main idea of the book is to present that there are a variety of memory techniques suited to every fancy. Their simplicity and time consuming stress their accessibility to each person in any age. If you think, that people who can perform incredible mnemonic feats for example, as remembering hundreds of random numbers or the exact order of a shuffled deck in a minute, are unique, you bark up the long tree. They also do not have the photographic memory, their brains are the same like yours. Memorizing lots of random cards and digits are based on the fundamental techniques. The ultimate message is that by coding the essential information using vivid, picturesque mental images, you can reliably code both information and its structure. And because the images are vivid (as noticed in the majority of the memory techniques), they are easy to recall when you need them.

The most effective memorizing strategies can help you to improve your efficiency with a variety of tasks!

Having discovered these techniques and then mastered them, you are going to be able rapidly and easily memorize and learn both complex and simple pieces of the lists and data. If you now finally realize that you wish to tap into a new kind of mind power and are „heads up" to create a better brain, which will open the door for the new levels of intellectual potential, then use the memory techniques provided in this book because it will force your results.'

Author's Afterthoughts

Thanks ever so much to each of my cherished readers for investing the time read this book!

I know you could have picked from many other books but you chose this one. So a big thanks for downloading this book and reading all way to the end.

If you enjoyed this book or received value from it, I'd like to ask you for a favor. Please take a few minutes to post an honest and heartfelt review on Amazon.com Your support does make a difference and to benefit other people.

Copyright 2017 by Karen Chapman - All rights reserved.

All rights Reserved. No part of this publication or the information in it may be quoted from or reproduced in any form by means such as printing, scanning, photocopying or otherwise without prior written permission of the copyright holder.

Disclaimer and Terms of Use: Effort has been made to ensure that the information in this book is accurate and complete, however, the author and the publisher do not warrant the accuracy of the information, text and graphics contained within the book due to the rapidly changing nature of science, research, known and unknown facts and internet. The Author and the publisher do not hold any responsibility for errors, omissions or contrary interpretation of the subject matter herein. This book is presented solely for motivational and informational purposes only.